SRA
Connecting Math Concepts

Level A Workbook 1

COMPREHENSIVE EDITION

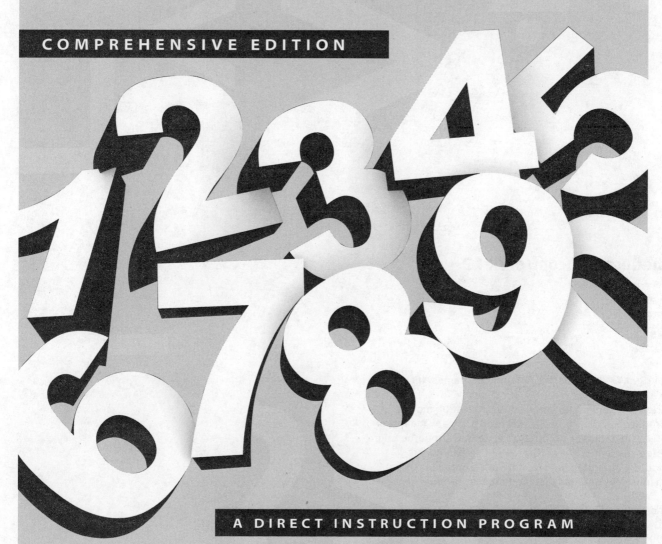

A DIRECT INSTRUCTION PROGRAM

 **Education**

Bothell, WA • Chicago, IL • Columbus, OH • New York, NY

mheducation.com/prek-12

Copyright © 2012 The McGraw-Hill Companies, Inc.

Send all inquiries to:
McGraw-Hill Education
4400 Easton Commons
Columbus, OH 43219

ISBN: 978-0-02-103572-4
MHID: 0-02-103572-5

Printed in the United States of America.

20 21 LON 25 24 23

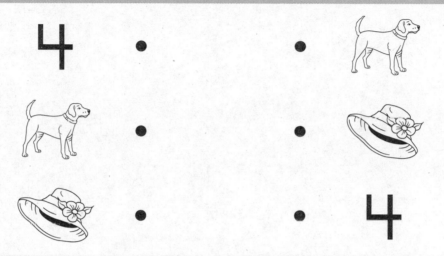

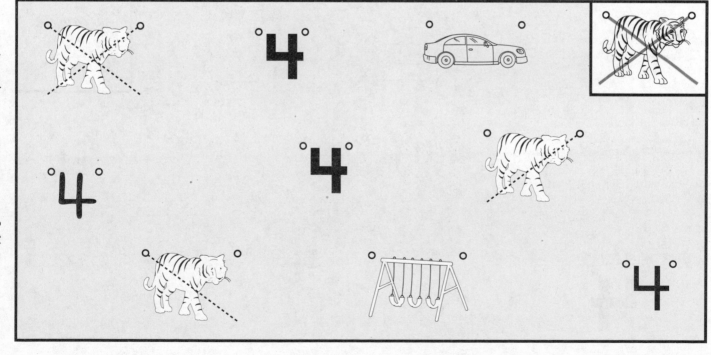

Connecting Math Concepts

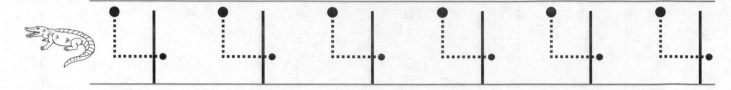

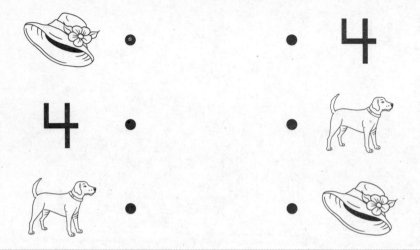

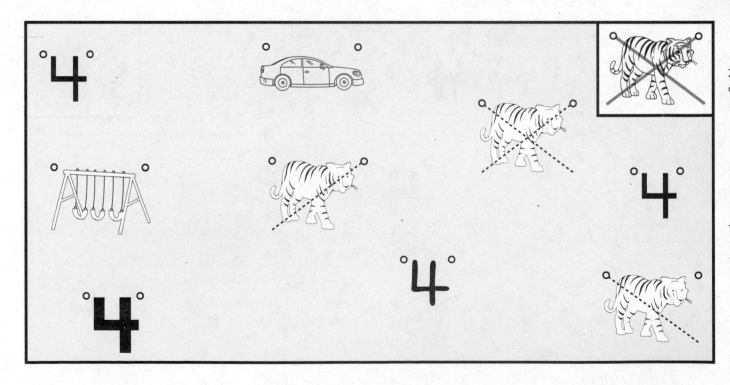

Connecting Math Concepts

Name _____

2

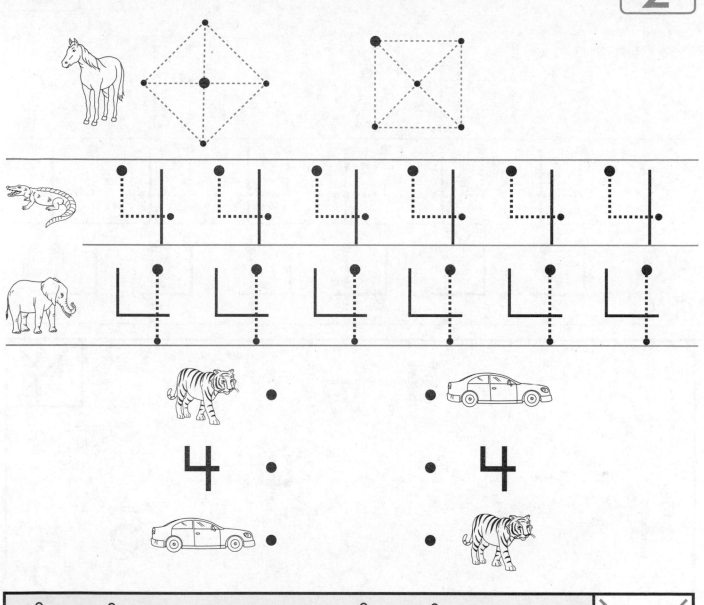

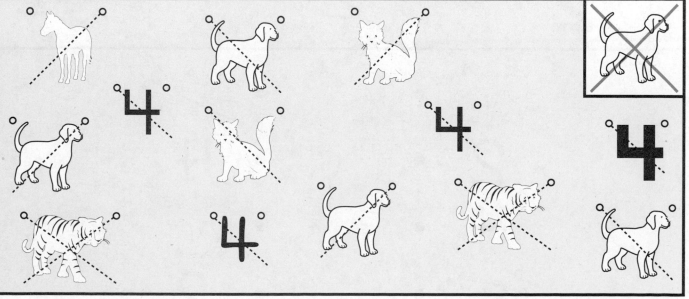

Connecting Math Concepts

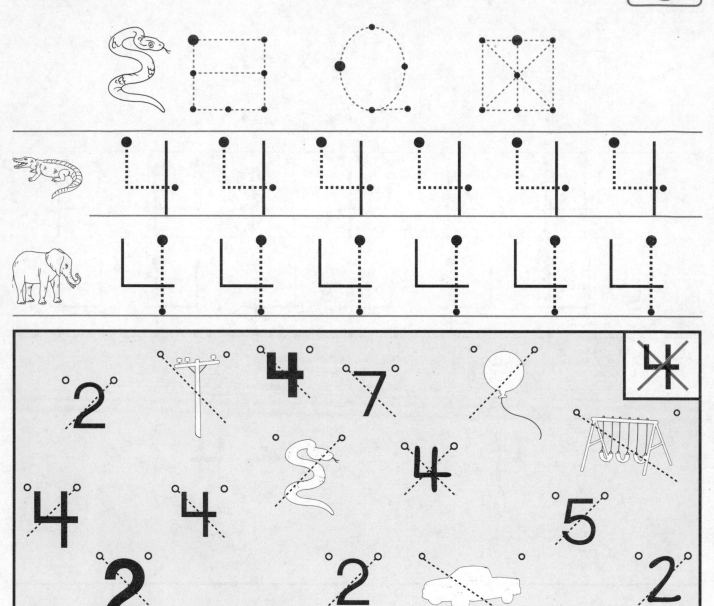

Connecting Math Concepts

Name _____

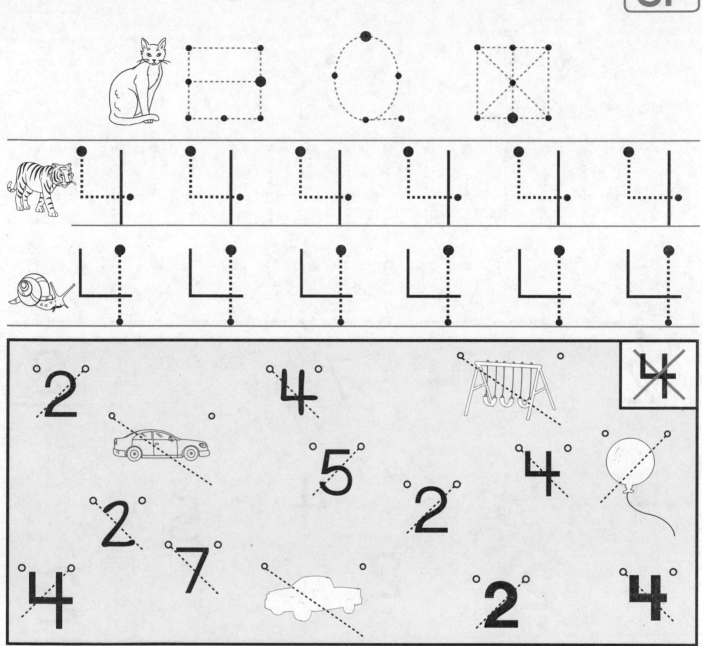

4

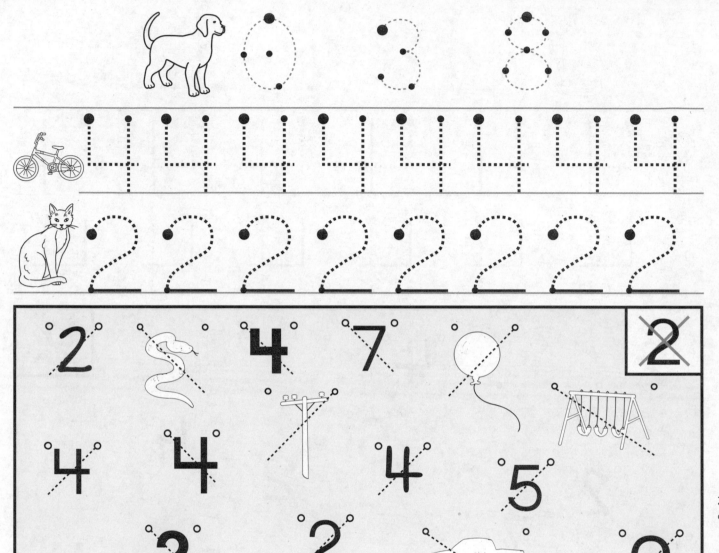

Connecting Math Concepts

5

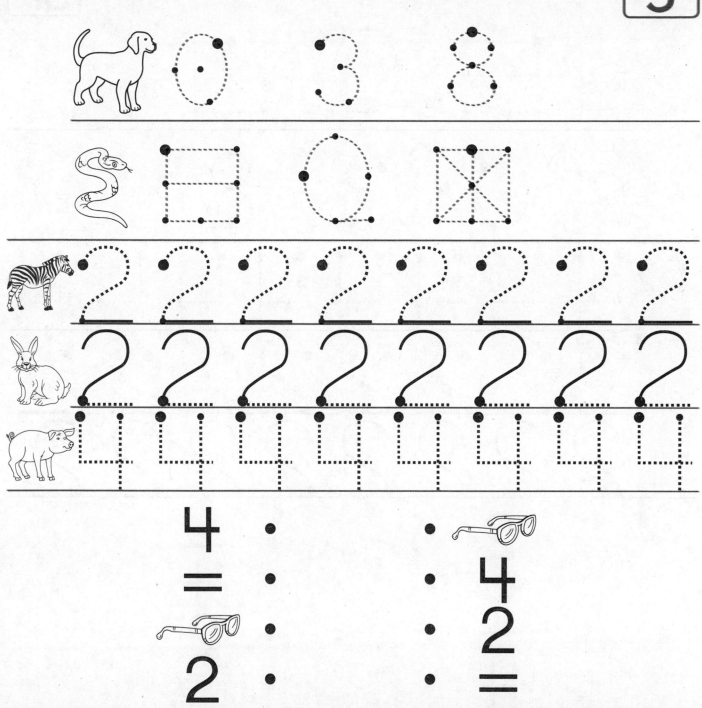

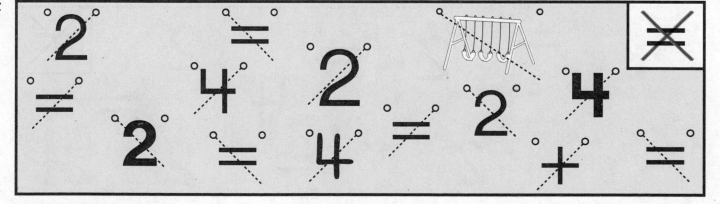

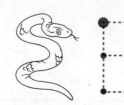

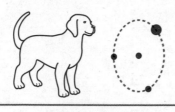 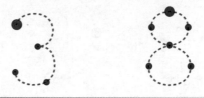 0 3 8

 4 4 4 4 4 4 4 4

 2 2 2 2 2 2 2 2

2 2 2 2 2 2 2 2

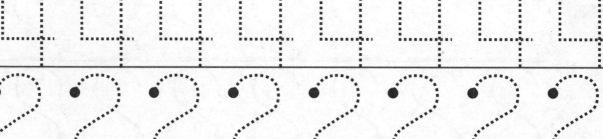

2
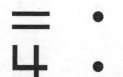
=
4

:
•
:
•
:
•
:
•

=
4
2
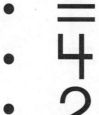

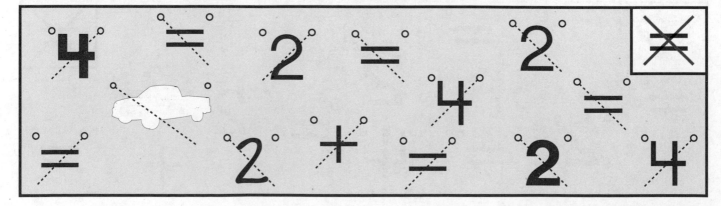

5P

Connecting Math Concepts

Name

Name _____

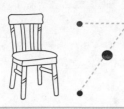 7   ⊠   3

 6   2   ⊞

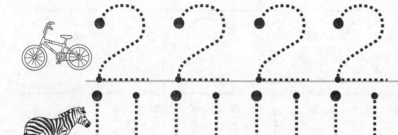 2 2 2 2 2 2 2 2

4 4 4 4 4 4 4 4

= •   • ☐   = •   • 2

4 •   • =   4 •   • 4

☐ •   • **2**   **2** •   =

2 •   • **4**   ☐ •   • ☐

6   4   =

2   6   4   2

4   =

   4   4   2

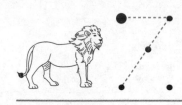 7 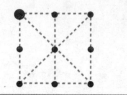 3

 6 2

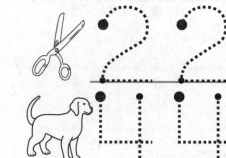 2 2 2 2 2 2 2 2

4 4 4 4 4 4 4 4

4 •
= •
2 •
☐ •

• =
• ☐
• 4
• 2

= •
2 •
4 •
☐ •

• 4
• 2
• ☐
• =

6 =
2
= 4

6

4
= 6
2 =

 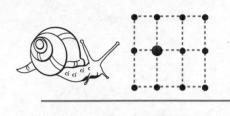 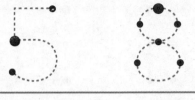

5   8

3   9

2 2 2 2 2 2 2 2

4 4 4 4 4 4 4 4

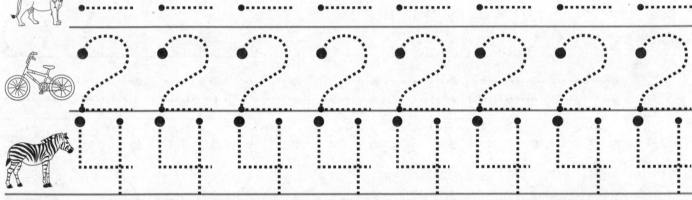

6   4   =   2   ~~6~~

☐   2   4   ☐

2   **6**   =   6

4 •   • 6   2 •   • 2

6 •   =   • 2   6 •   • ☐

= •   • 2   ☐ •   • 4

2 •   • 4   4 •   • 6

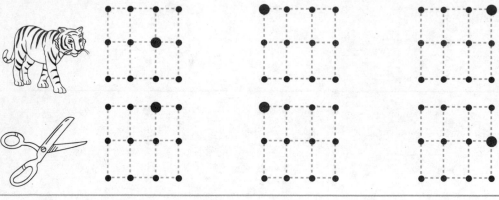

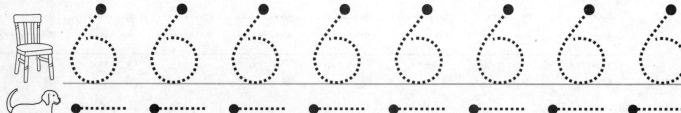

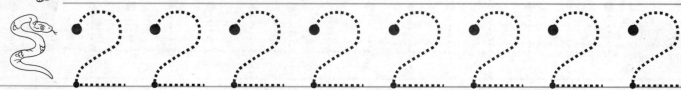

2 •   • 6   = •   • 6

4 •   • ☐   6 •   • =

6 •   • 2   2 •   4

☐ •   • 4   4 •   • 2

☐   7   2   6   2̶

4

2   7   2   4   2   ☐

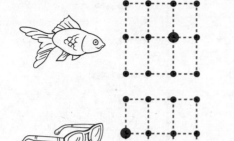

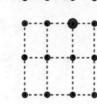

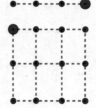

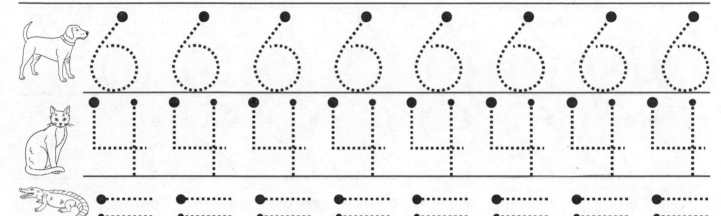

6 6 6 6 6 6 6 6

4 4 4 4 4 4 4 4

6 •   • 4 | 4 •   • 6

= •   • 2 | 6 •   □

2 •   • 6 | □ •   • 2

4 •   • = | 2 •   • 4

7 = 6   2   6 ⊘

+ 2 4   =   +

4   =   4   6

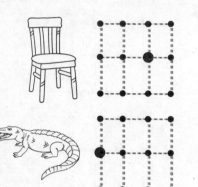
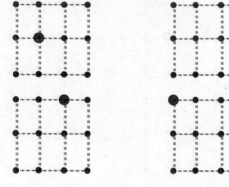

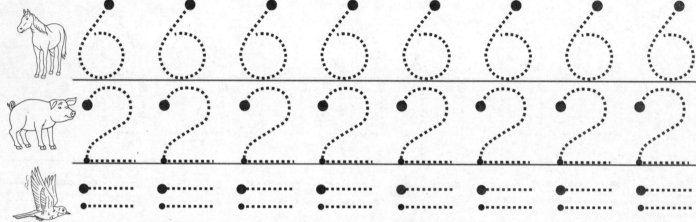

6 •   • 4   4 •   • 6
= •   • 2   □ •   • □
4 •   • 6   6 •   • 2
2 •   = •   2 •   • 4

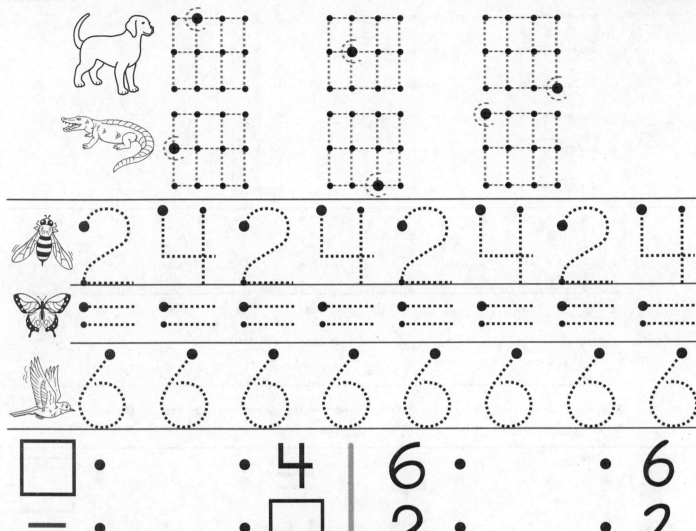

Name _____

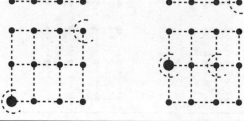

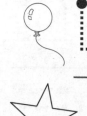

4 2 4 2 4 2 4 2 4

6 6 6 6 6 6 6 6 6

7 •   • 4   = •   • □
2 •   • 7   4 •   • =
6 •   • 2   □ •   • 7
4 •   • 6   7 •   • 4

= 2 4 = 7 ✕

7 □ = □ 2 □ =

Copyright © The McGraw-Hill Companies, Inc.

Connecting Math Concepts

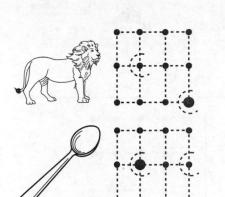

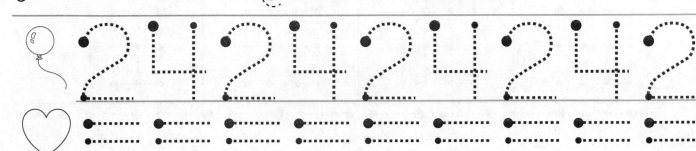

4 •   • 7   4 •   • ☐
2 •   • 4   ☐ •   • 4
6 •   • 2   = •   6
7 •   • 6   6 •   =

= 2   4 = 6   7 = [X̶]
7 ☐   7 ☐   2 ☐ =

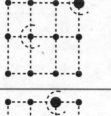 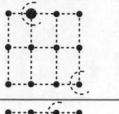 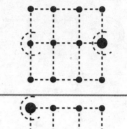

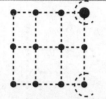 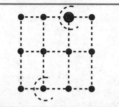 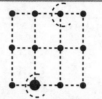

7 7 7 7 7 7 7 7 7

62 = 62 = 62 = 6

4 4 4 4 4 4 4 4 4

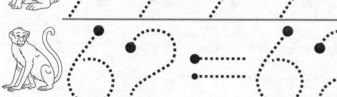

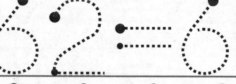 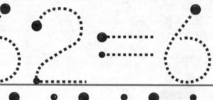

```
6 •        • =      4 •        • 7
7 •        • 6      7 •        • 4
= •        • 7      2 •        • 2
```

```
7   5   2   7   6   2
   4     7     5   3   2   7
```
(with the last 7 crossed out in a box)

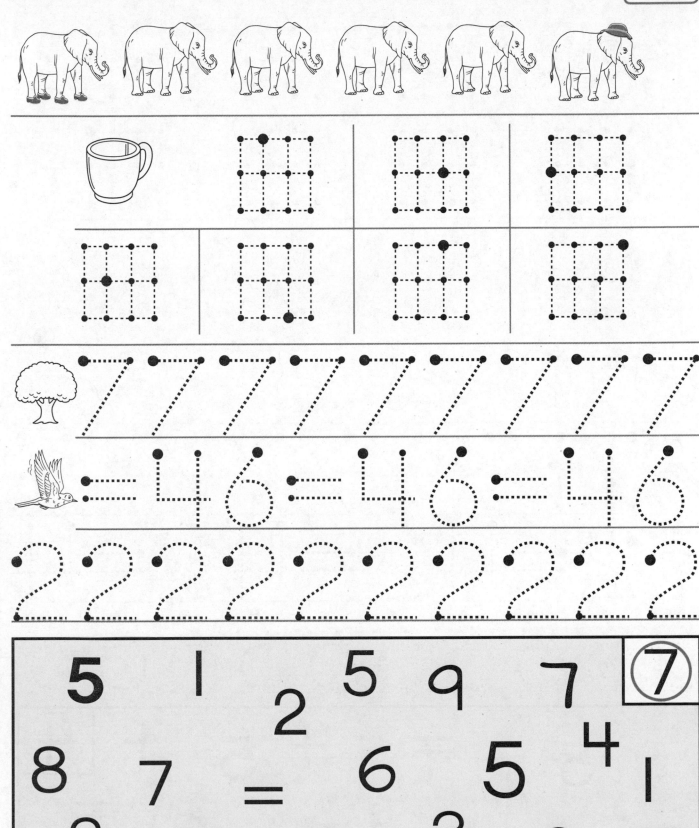

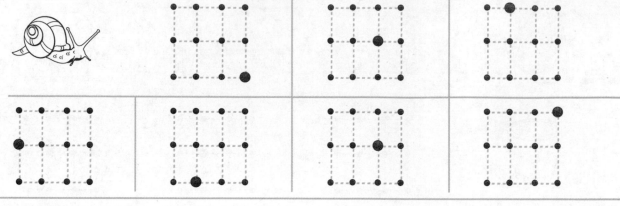

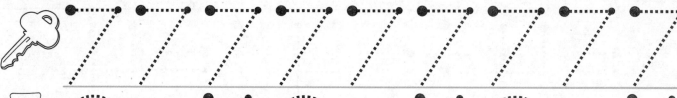

7 7 7 7 7 7 7 7 7

2 = 4 2 = 4 2 = 4

6 6 6 6 6 6 6 6 6

4

5   1   =   4   5   +   4

=   +   9   4   −   9   =   2

5 5 5 5 5 5 5 5 5

 6 7 2 6 7 2 6 7 2

5 + = 9 7

3 1 2 4 6 5 =

5

7 5 4 5 + = 9

+ = 4

+ • • □ 4 • • 6

= • • 7 5 • • 4

□ • • + 6 • • 7

7 • • = 7 • • 5

Name _____

15P

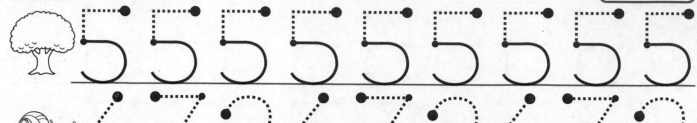

5 5 5 5 5 5 5 5 5 5 5

6 7 2 6 7 2 6 7 2 6 7 2

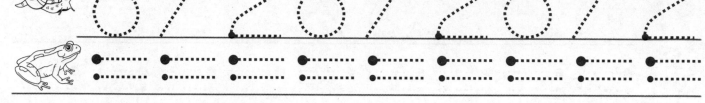

5   +   =   +   7   5̶   ⊜

4   =   5   4   6   5   +

   2

7   =   4   1   =   2

         +   =   4

4 •	• =	6 •	• 7
= •	• 4	2 •	• ☐
+ •	• 5	☐ •	• 6
5 •	• +	7 •	• 2

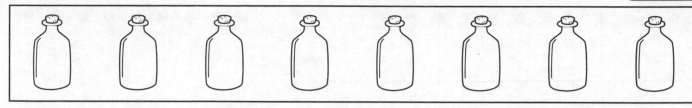

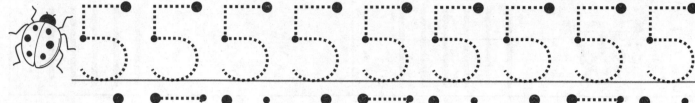

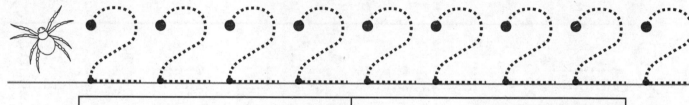

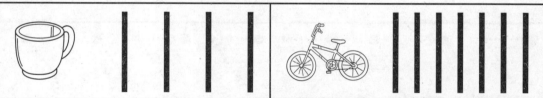

7 •   • 2

2 •   • 5

5 •   • 7

4 •   • 4

2 + ~~5~~ Ⓞ=Ⓞ

6

+ = 7   **5**

5

2   9 =

5   =   3

+

 • • • • • • • • •

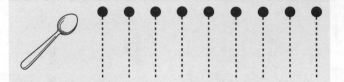 • • • • • • • • •

 | | | | | | | | |

 | | | | | | | | |

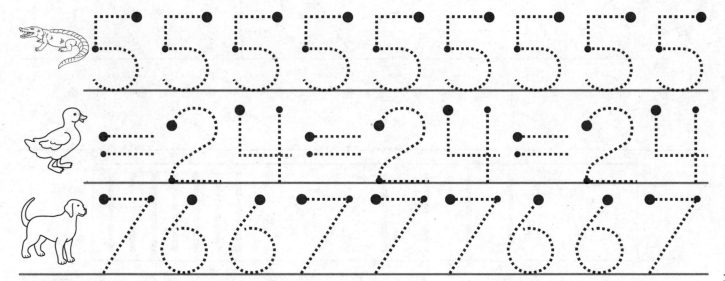

55 5 5 5 5 5 5 5

=24=24=24

767 7 7 7 7 6 7

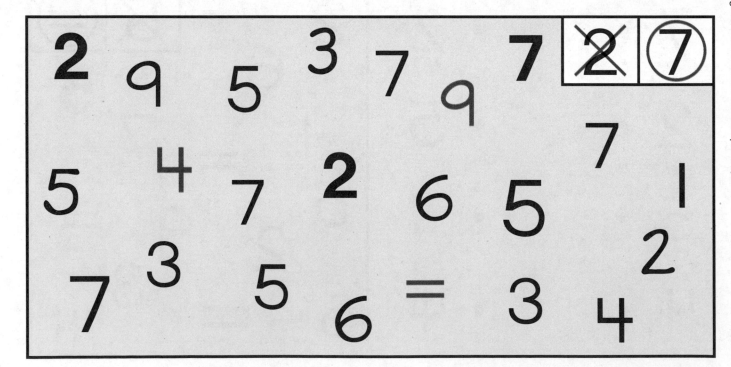

Connecting Math Concepts

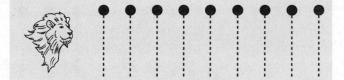

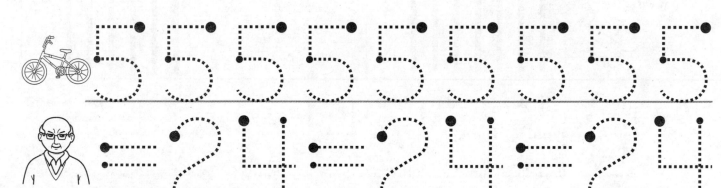

5 5 5 5 5 5 5 5 5 5

2 4 2 4 2 4

7 6 6 7 7 7 7 6 6 7

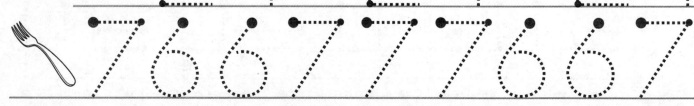

2 9 5 7 9 7 ☒ ⑤
4 3
5 7 2 7 1
6 5
7 3 5
5 6 = 3 4 2

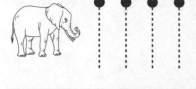

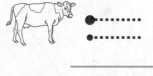

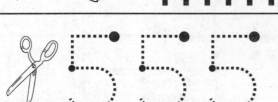

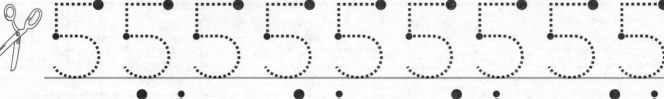

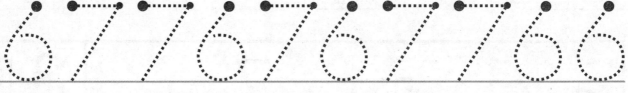

5 9 = 3 7 9 5   ≠   ⑤

3 4 5 = 2 3 =

4 7 3 = 2 5

Connecting Math Concepts

Name ____________________

19

+ + + + + + + + +

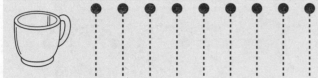

5 5 5 5 5 5 5 5 5

7 = 2 7 = 2 7 = 2

4	3

2 + 3 5 4 9 2 3 9 4 5 = 3 3 3 4 5 + 7 4

+	=	4	5
=	3	5	6
2	+	6	7
3	2	7	4

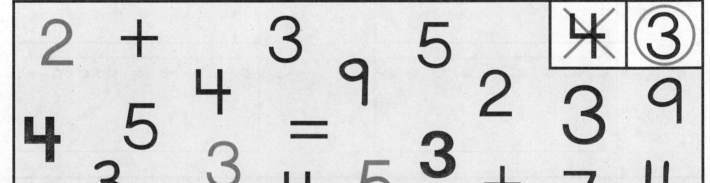

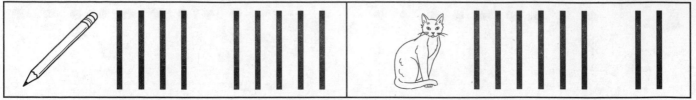

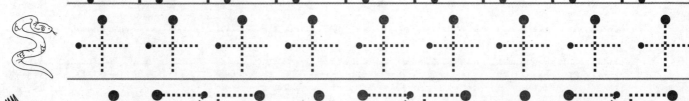

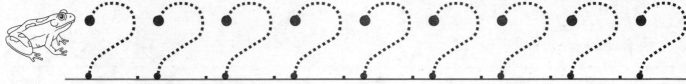

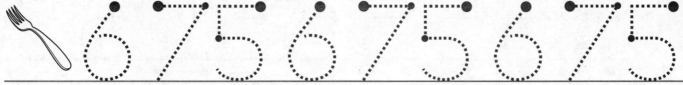

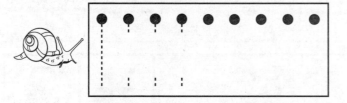

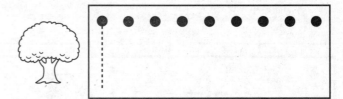

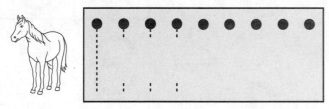

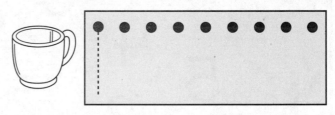

3   7   7   4   =   X̶   ②

2   5   2   3   5   2

2   =   7   +   3   2   3

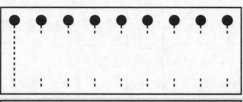

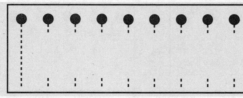

‖‖‖ ‖‖ | | | | ‖‖‖

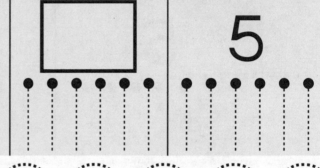

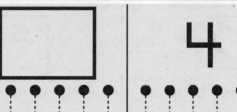

4   5

2 2 2 2 2 2 2 2 2 2

+ = + = + = + = + = +

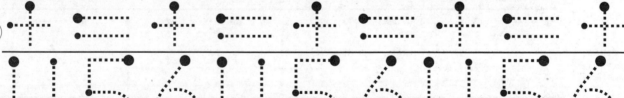
4 5 6 4 5 6 4 5 6

1
+
7
=

+
=
1
7

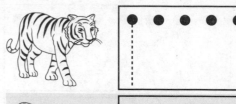
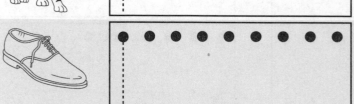
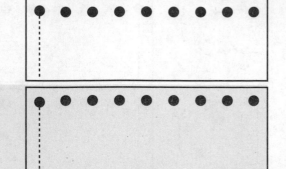

5   2   □

2 2 2 2 2 2 2 2 2 2

7 4 7 4 7 4 7 4 7 4

5 5 5 5 5 5

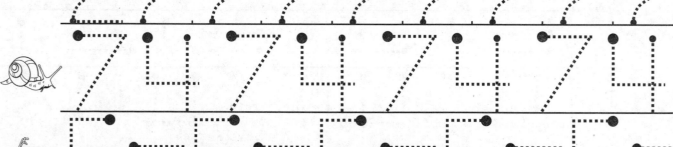

3 = **3**   + 5   □   ☒ ⊕

2 + 3 = 6   9   +

Connecting Math Concepts

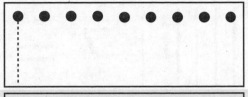

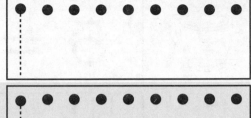

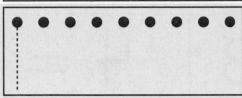

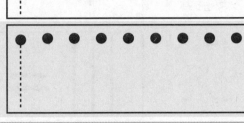

3

6

2 2 2 2 2 2 2 2 2

7 5 7 5 7 5 7 5 7 5

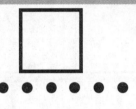

+ = + 3 5 ☐ ⊁ ⊕

2 3
+ = 3 6 4 9 +

5 = ||||| 7 = |||||||

|||| = 3   4 = |||||

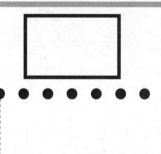

6   [ ]   2

3 3 3 3 3 3 3 3 3

| | | | | |

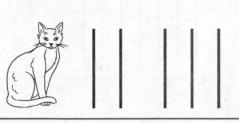 $\boxed{}$ 6 | | |

 5
• • • • • •

1
• • • • • • •

7
• • • • • •

 ||||||| = 7   5 = ||||| |||| = 4

 4 4 4 4 4 4 4 4 4 4

5 5 5 5 5 5 5 5 5 5

 6 3 6 3 6 3 6 3 6 3

2 2 2 2 2 2 2 2 2 2

$1 \quad 4 \quad 4^4 \quad 1 \quad + \quad 5 \quad 7 \quad 2 \quad \cancel{2} \quad \boxed{1}$
$3 \quad 2 \quad 1 \quad 7 \quad + \quad = \quad 2 \quad 7 \quad 3 \quad 1 \quad 5$

25

3   4

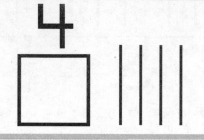

$5 = |||||$ $||||||||| = 9$ $|||| = 4$

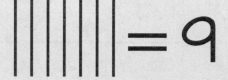
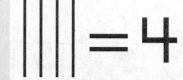

3   5   ☐   4

• • • • •   • • • • •   • • • • •   • • • • •

3 3 3 3 3 3 3 3 3

4 4 4 4 4 4 4 4 4

5 5 5 5 5 5 5 5 5

7 7 7 7 7 7 7 7 7

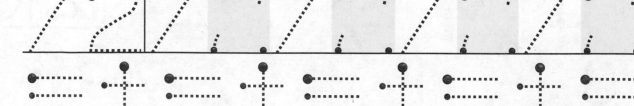

Connecting Math Concepts

|||| |||| ||||

 6 □ 3
· · · · · ·   · · · · · ·   · · · · · ·

 3 + 4
· · · · · · ·

  5 + 2
· · · · · · ·

  |||| = 4 6 = ||||| |||||| = 7

 3 + 4
||||

5 + 2
|||

 4 L L L L L L L L L L L

3 3 3 3 3 3 3 3 3 3

5 5 5 5 5 5 5 5 5 5

+ · = · · · · · · · · · · · · · · · · ·

4 = •••••••••   ••••••=5   2 = •••••

 ||| ||||||||| |||||

 5 + 3   7 + 2
•••••••   ••••••••

🍴 7 = ||||||| 5 = ||||| ||||=6

🚲

 5 2

9 2 5   1 8 7 ⨉6 ⑨
3 6 6 5 9 6 9 6 4
  1 8 2 5 6 2 9

Connecting Math Concepts

Copyright © The McGraw-Hill Companies, Inc.

Name _____

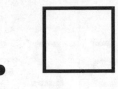 28

 • ☐ | • 6 | • 5

 • • • • • • = 4 | 2 = •

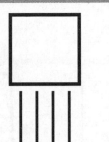 ☐ ☐ ☐
|||| || |||||

 9 + .2   4 + .3

 |||||||| = 6   7 = |||||| 9 = ||||||

4 4 4 4 4 4 4 4 4

5 5 5 5 5 5 5 5 5 5

2 3 / 3 / 3 / 3 / 3 7

2 + . = .   ? . .   ? .   ? . .

Name _____

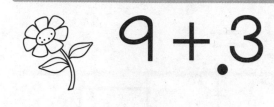 3 = ••••• • = 5   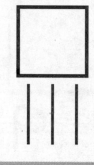 2 = •

 9 + .3   10 + .4

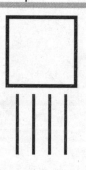

 |||||||| = 9   5 = |||||| 7 = |||||||

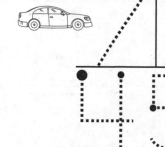 7

45 ⌐ ⌐ ⌐ ⌐ ⌐ ⌐

3 6 3 6 3 6 3 6 3 6 3 6

| + = |

Connecting Math Concepts

 • = 5 | 7 = •

 □   □   □
||||| |||| |||||

 3 + 4 | 6 + 2

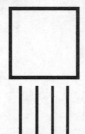 |||| = 4 | 9 = |||||||| | 3 = |||

 9 9 9 9 9 9 9 9 9 9 9

7 4 7 7 7 7

5 2 2 2 2 2 2 2

3 6 3 6 3 6 3 6 3 6 3 6

6 = •   • = 4   5 = •

 ☐ | | | | | | | | | | ☐ | | ☐ | | | | | | | | |

 | | | | = 5 | | | | | | | = 7   3 = | | | |

 3 + •5   7 + 4

 9 9 9 9 9 9 9 9 9 9 9 9

0 0 0 0 0 0 0 0 0 0 0

3 3 3 3 3 3 3 3 3 3

7 7 7 7 7 7 7 7

☐ .6   ☐ .3

||||| ||||||

9 + 4

||||| = 4   6 = ||||| ||||| = 5

|| = 3 ||||||| = 7   9 = |||||||||

 0 9 0 9 0 9 0 9 0 9 0 9 9

3 6 3 6 3 6 3 6 3 6 3 6 3 6 3

4 1

5 7

+ =

2 + .4 | 8 + .3

    •   = 82 = •   •   = 3

    ☐   .5   ☐   .1

|||| ||||||

    4 = |||||| |||| = 5 |||| = 3

|||||||||| = 9   8 = |||||||||||| 6 = ||||||

 0 9 0 9 0 9 0 9 0 9 0 9 9

7 1

5 2

Connecting Math Concepts

. 6

| | | | | |

.5

 5 + .2 =   6 + .3 =

 7 = .   .   = 4

 9 = ||||||||| ||||| = 5 |||||| = 6

|||||||| = 8   7 = ||||||| 4 = |||

 4

+ =

0 3 0 3 0 3 0 3 0 3 0 3

2 5 2 2 2 2

_____

=6 | 2= | =3

||||||| = 7 | 3 = ||| | |||| = 4

| = || | ||||||||| = 9 | ||| = 2

.5 | [ ] | [ ] | [ ]
| |||| | || | |||||

4 . . . .

7 . . . .

9 0 9 0 0 9 0 9 0

_____ 6 _____ 7 _____ 8 _____ 9

 _____

 7 = •   • = 2   • = 4

 |||||||| = 8 |||| = 3   5 = |||||

||||||| = 6   9 = ||||||||| ||||||| = 7

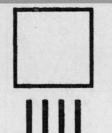 ☐   • 3   ☐   • 6

      ||||| ||||

 3 ? ? ? ? ? ? ? ? ?

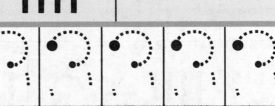 0 9 6 0 9 6 0 9 6 0

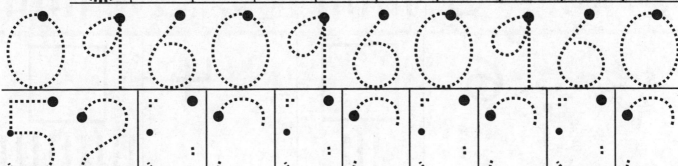 5 2 ? ? ? ? ? ? ? ?

|||||·| =

 8   9   6   4
___   ___   ___   ___

 ·
_____

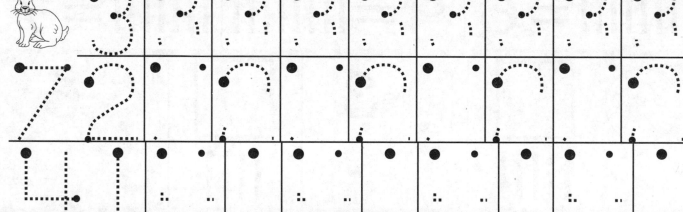

|||||||| = · | · = |||| | · = |||||

/ |||| = 3 ||||||| = 7   5 = ||||||

 · 6   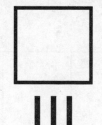[ ]   · 4   [ ]

Name ____________________

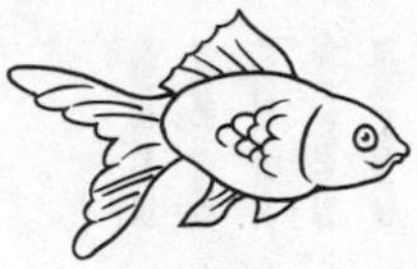

|||||| • | ||||| •

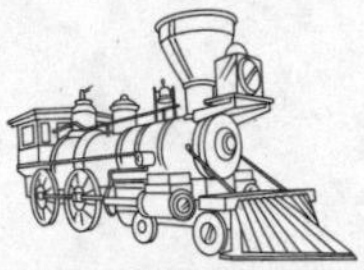

6 9 5

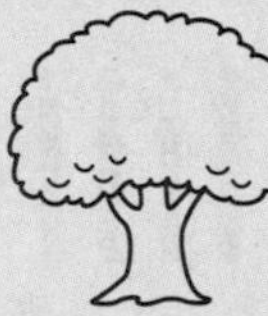

|||||||| • =

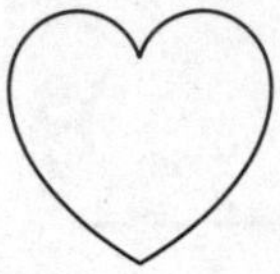

• ______ | 3 + 2 ______

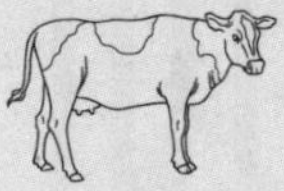

• =|| | |||||||||= • | • =|||||||

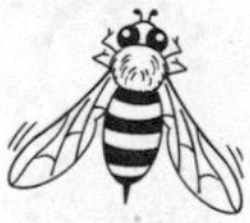

5 2

3 6 3 6 3 6 3 6 3 6

7 9 9 9 9 9

+ =

• =7 | • =8 | 3= •

| | | | | • | | | | | | | • | | | | | | |

7   5   9   4

| | | | | | | • = | | | | | • =

•
_____   6 + 1

| | | = •   • = | | | | | | • = | | | |

8 8 8 8 8 8 8 8 8 8

0 6 6 6 6 6

5 2

= +

5 = •   • = 4 1 = •

Connecting Math Concepts

|||||||| • = |||||| =

   8   5   7   4   9

 _____ | _____

 9 0 0 0 0 0 0 0 0 0 0

8 8 8 8 8 8 8 8 8 8 8

3 5 ? ? ? ? ? ? ? ? ?

 • = ||||| |||||||| = •   • = ||

 = 6   = 3   1 =

• = 4   2 = •   5 = •

Name _____

 ||||||||| = ˙ | ||||||| = ˙

 3   7   4   8   5

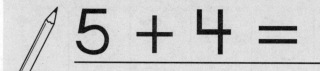 5 + 4 = ˙ | 5 + 2 = ˙

||||||||||| = ˙˙ | ˙˙ = |||||||| ˙ = |||||||

9 9 9 9 9 9 9 9 9 9 9

8 8 8 8 8 8 8 8 8 8 8 8

4 4 4 4 4 4 4 4 4 4 4

3 5 2 2 2 2 2 2 2 2 2

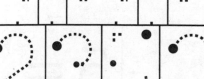
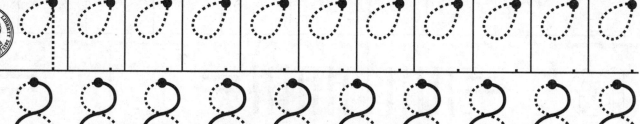

Connecting Math Concepts

•
_____

 ||||| = • ||||||| = •

 5   9   3   6   4

6 + 1 = •   6 + 3 = •
_____   _____

 ||||| = •   • = |||||||| |||||| = •

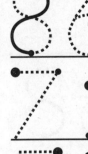

0 9 ( ) ( ) ( ) ( ) ( ) ( ) ( ) ( )

8 6 8 6 8 6 8 6 8 6 8 6

7 2))))))

5 3

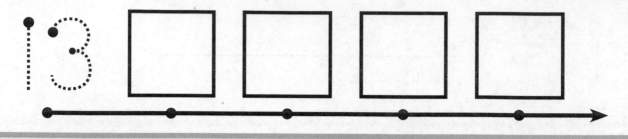

 1 + 4 = ˙ ___   3 + 4 = ˙ ___

||||| = ˙ ||||||||| = ˙

 7 + 2 = ˙ ___

 ||||||| = ˙   ˙ = ||||| ˙ = ||||||||||

 0 0 0 0 0 0 0 0 0 0

8 8 8 8 8 8 8 8 8 8 8

35 2 2 2 2 2 2 2 2 2

Connecting Math Concepts

$5 + 3 =$

♡ ☐ $- 4 =$ ☐ $- 2 =$

$2 + 5 =$ $2 + 4 =$

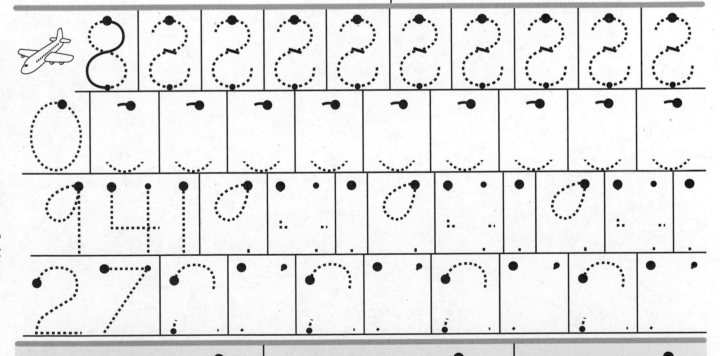

Name _____

3 + 2 =

□ − 4 =   □ − 3 =

7 + 2 = ____   6 + 1 = ____

☆ 8 2 2 2 2 2 2 2 2 2

3 4 3 3 3 3 3 3 3 3

0 2 2 2 2 2 2 2 2 2

9 5 0 0 0 0 0 0 0 0

+ = ...

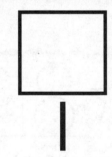
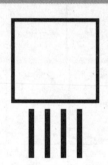

□ − 5 = □

□ − 4 = □

|||||| |||||||

|||||||||| = •   • = ||||| |||||| = •

||||||| = •   • = |||||||||| ||||||| = •

3   [□ over ||||]   6   [□ over |]

|||||| = 7 ||||| = 5   9 = |||||||||

8 = ||||||| 1 = 1 |||||||||||| = 12

Connecting Math Concepts

$$7 + 2 =$$

$$2 + 4 =$$

🍎 ☐ $- 4 =$

☐ $- 1 =$

🍴 | | | | |

🐘 $3 + 1 =$ _____

$5 + 2 =$ _____

♥ ||||| =

= |||||

|||||| =

■ $- 5 =$ ■

■ $- 3 =$ ■

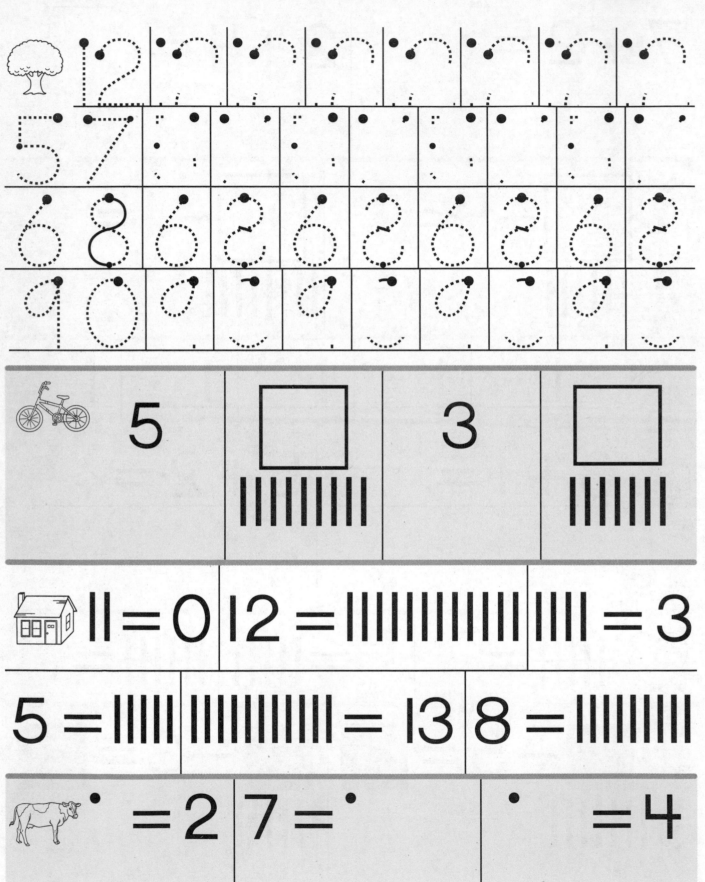

Connecting Math Concepts

6 + 3 = | 5 + 2 =

1 + 4 =

□ − 3 =   □ − 2 =

‖‖‖   ‖‖‖‖‖‖

□ − 5 =

‖‖‖‖‖

🌙 ||||||||||| = | = 卌

= |||||卌 ||||| =   = ||卌卌

🐱   3   ☐ |||||||| ☐ || 7

🏠 |||||卌 = 6   15 = ||||||||||||| = 0

13 = ||||||||||卌 ||卌 = 2 |||||卌 = 8

🕷   = 5   1 =   = 4

6 =   = 3   = 2

$\square - 4 =$ $\square - 2 =$

IIIIII   IIIIIII

$15 + 4 =$ $4 + 3 =$

 _____

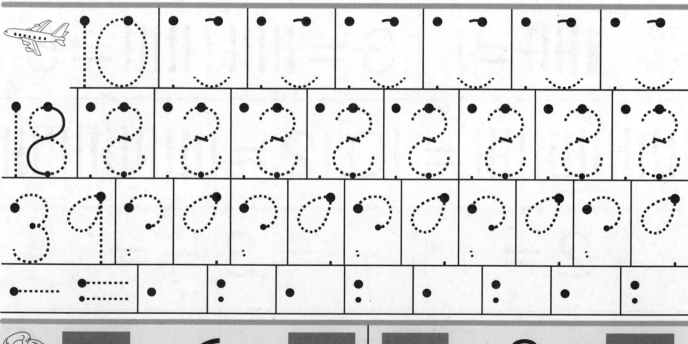

 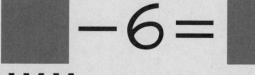 $\square - 6 =$ 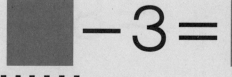 $\square - 3 =$

IIIIII   IIIIII

★ ||||| =   = ||||||||||||

||||||| =   = |||||||||| ||||| =

🐯 2   [▢ | ||||||||]   [▢ | ||||]   5

🚩 ||||| = 1   3 = |||||| ||||| = 5

|||||||||||||||| = 10   12 = ||||||||||||||||

🐄 2 =   = 3   4 =

= 8   5 =   = 6

Connecting Math Concepts

17 + 2 =

1 + 4 =

13 + 5 =

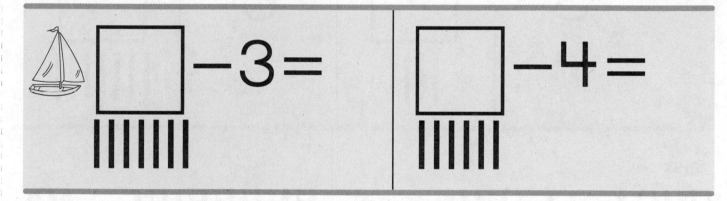

10

4 9

5 8

♥ |||||||| =   =|| ||||||||| =

|||||||||| = ||||||| = |||||||| =

3   □ (||||)   6   □ (||||||)

||||||| = 4 |||||| = 5 |||||||||||| = 10

6 = |||||||| |||||||||| = 8 ||||||||| = 9

= 7   = 6   2 =

8 =   = 0   = 4

Connecting Math Concepts

8 + 4 = | 15 + 2 =

□ − 1 =   □ − 5 =

_____   _____

☆ ‖‖‖ =   ‖‖‖‖‖ =   = ‖‖‖

‖‖ =   = ‖‖‖‖   ‖‖‖‖‖ =

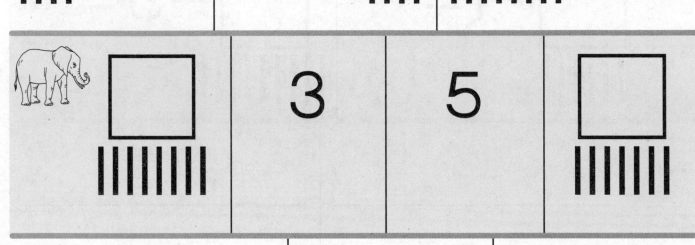

📱 ‖‖‖‖ = 3   5 = ‖‖‖‖‖   ‖‖ = 0

6 = ‖‖‖‖‖‖   ‖‖‖‖‖‖ = 8   ‖‖‖‖‖ = 4

🐟 1 =   = 2   3 =

= 4   5 =   6 =

Connecting Math Concepts

14 + 3 =   4 + 6 =

☐ − 4 =   ☐ − 5 =

4 + 1 =   8 + 1 =

3 + 1 =   5 + 1 =

_____

= ‖‖‖‖‖‖‖ ‖‖‖ ‖‖ =   = ‖‖‖‖‖‖‖ ‖‖

= ‖‖‖‖‖‖‖ ‖‖‖ ‖‖‖‖‖ ‖‖‖‖ =   ‖‖‖ =

| [ ] | 0 | [ ] | 4 |
| ‖‖‖‖‖ | | ‖‖‖‖‖ | |

‖‖‖‖‖‖‖‖‖‖ = 10   8 = ‖‖‖‖‖‖‖‖   ‖‖‖‖‖ = 0

‖‖‖ = 2   3 = ‖‖‖‖‖‖   ‖‖‖‖‖‖‖‖‖‖ = 12   ‖‖‖‖ = 4

= 5   4 =   = 3

= 10   = 6   2 =

Connecting Math Concepts

$3 + 5 =$ $17 + 2 =$

_____ _____

🐁 $7 - 3 =$ $6 - 5 =$

☆ $6 + 1 =$ $2 + 1 =$

$7 + 1 =$ $5 + 1 =$

 _____   _____

9 0 0 0 0 0 0 0 0

8 2 2 2 2 2 2 2 2

7 3 3 3 3 3 3 3

$$||||||||||||||||| = \qquad \qquad = |||||$$

$$= ||||||||||| \qquad ||||||||| = \qquad |||||||| =$$

3   ☐   ☐   7
       | |||||

$$||||||||| = 10 \qquad 12 = ||||||||||||$$

$$|||||| = 4 \qquad ||||||| = 5 \qquad 9 = ||||||||||$$

$$||||||||||| = 70 = ||||||||||| = 0 \quad || = 3$$

_____ | _____

 ____ ____ ____ ____

$17 + 1 =$ | $10 + 1 =$

$4 + 1 =$ | $8 + 1 =$

 $8 - 3 =$ | $5 - 4 =$

$16 + 3 =$ | $2 + 5 =$

🏚 $= ||||| \bcancel{||||}$ $\bcancel{||||} ||| =$ $||||||| \bcancel{||||} =$

$|||||||||| =$ $||||||||| \bcancel{||||} =$ $| \bcancel{||||} =$

🚩 4   ▢ $||||| |$ ▢ $| |||||||$ 8

⭐ $|||||| \bcancel{||||} = 6$ $4 = |||||||| \bcancel{||||} | \bcancel{||||} = 0$

$8 = ||||||||| \bcancel{||||} = 2$ $3 = |\bcancel{||||} ||||| \bcancel{||||} = 7$

$|||||||||| \bcancel{||||} = 13$ $|||||||||| \bcancel{||||} = 12$

🌳 $= 3$ $5 =$ $= 0$

$10 =$ $= 2$ $= 4$

Connecting Math Concepts

$6 - 1 =$ | $5 - 3 =$

$16 + 1 =$ | $5 + 1 =$ | $2 + 1 =$

$9 + 1 =$ | $10 + 1 =$

_____ | _____

$13 + 4 =$ | $9 + 5 =$

$2 + 3 =$

$\boxed{} - 4 =$ | $\boxed{} - 1 =$

|||| | ||||||

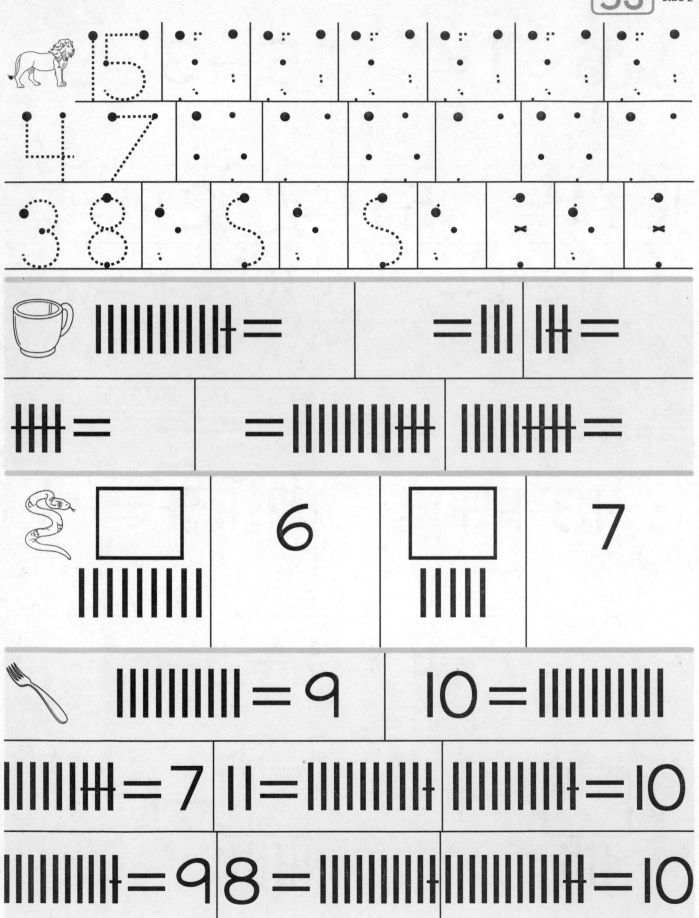

14+1=   7+1=   5+1=

9+1=   16+1=   3+1=

  _____ _____ _____

6 – 2 =   8 – 5 =

_____   _____

☐ – 3 =   ☐ – 4 =

||||||| |||||||

13 + 4 =   7 + 3 =

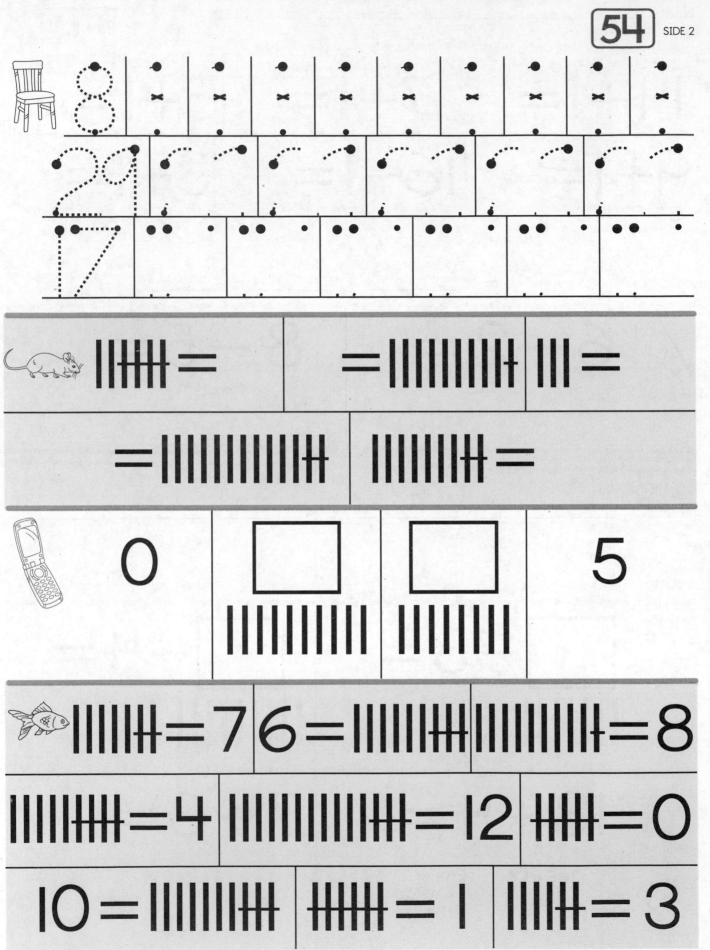

 _____

 _____ _____ _____ _____ _____

_____

| $13 + 1 =$ | $9 + 1 =$ |
| $12 + 1 =$ | $8 + 1 =$ |

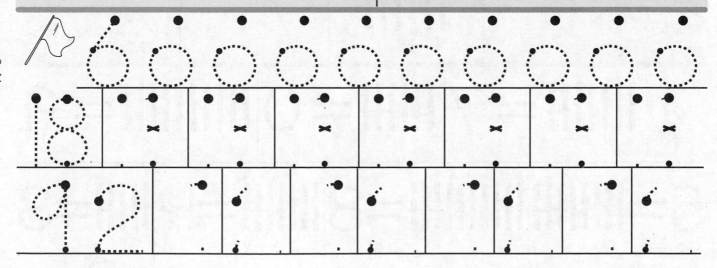

🔨 ☐ − 5 =   ☐ − 2 =

|||||| ||||||||

🐘 14 + 2 =   8 + 3 =

👞 |||||| =   = |||||| ||||| =

= ||||||||||| |||||| = ||||||| =

🕷 3   ☐   2   8

||||||

⛵ |||||| = 7 |||||| = 0 |||||||||| = 9

5 = ||||||||||||| = 8 |||||| = 4 |||||| = 3

8 + 1 =	10 + 1 =
17 + 1 =	6 + 1 =

_____ _____ — — — — —

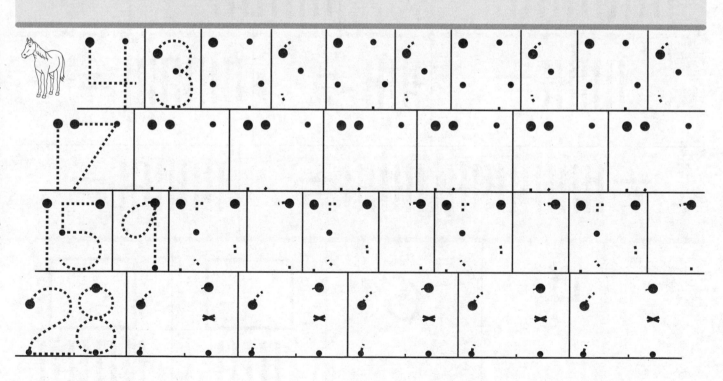

5 – 3 =　　7 – 1 =

 14 + 3 =　　15 + 6 =

▢ – 4 =

▢ – 2 =

|||||||||||||||

|||||||||

 ||||||| =　　||||| =　　||||||||| =

= |||||||||||||　　||||||||| =　　||||||||||| =

 4　　6　　▢　　▢

||||||　　|||||||

Connecting Math Concepts

_____

 _____ _____ _____

$$11 + 1 =$$

$$15 + 1 =$$

$$12 + 1 =$$

$$13 + 1 =$$

 _____

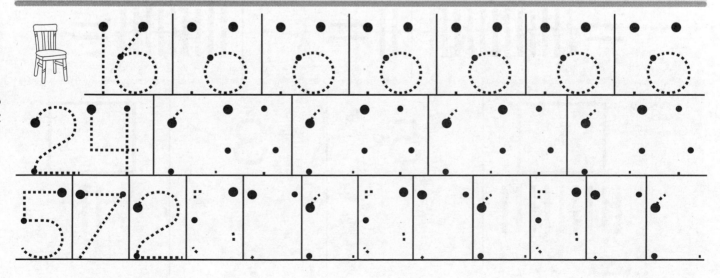

$\square - 4 =$

||||||||

$\square - 1 =$

|||||||

$11 + 3 =$

$6 + 4 =$

$7 + 2 =$

|||||||| =

||||||| =

||||| =

= |||||||||| ||| =

|||||||| =

 $\square$ 5 3 $\square$

|||||||||

||

Connecting Math Concepts

Name _____

 7 − 5 =   7 + 5 =

9 + 4 =   9 − 4 =

17 + 1 =   9 + 1 =

25 + 1 =   8 + 1 =

_____

15 + ▢ =   4 + ▢ =
    ‖‖‖   ‖‖‖‖

_____

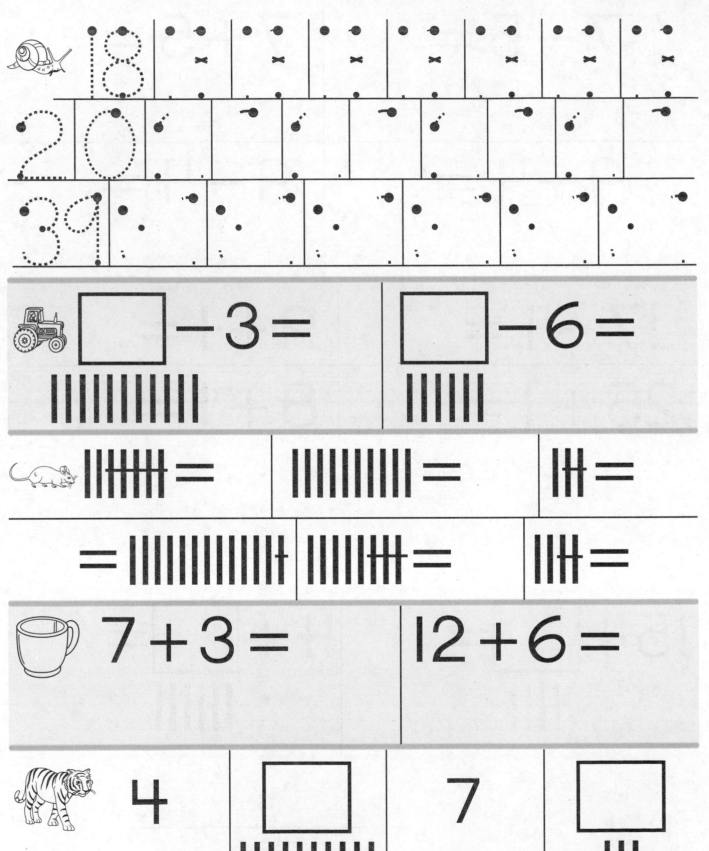

$$\boxed{} - 3 =$$

$$\boxed{} - 6 =$$

$$7 + 3 =$$

$$12 + 6 =$$

4

7

🐟 12 + 3 = | 6 − 5 =

35 + ☐ =
||||

 _____

🐰 _____

8 + 1 = | 3 + 1 = | 9 + 1 =

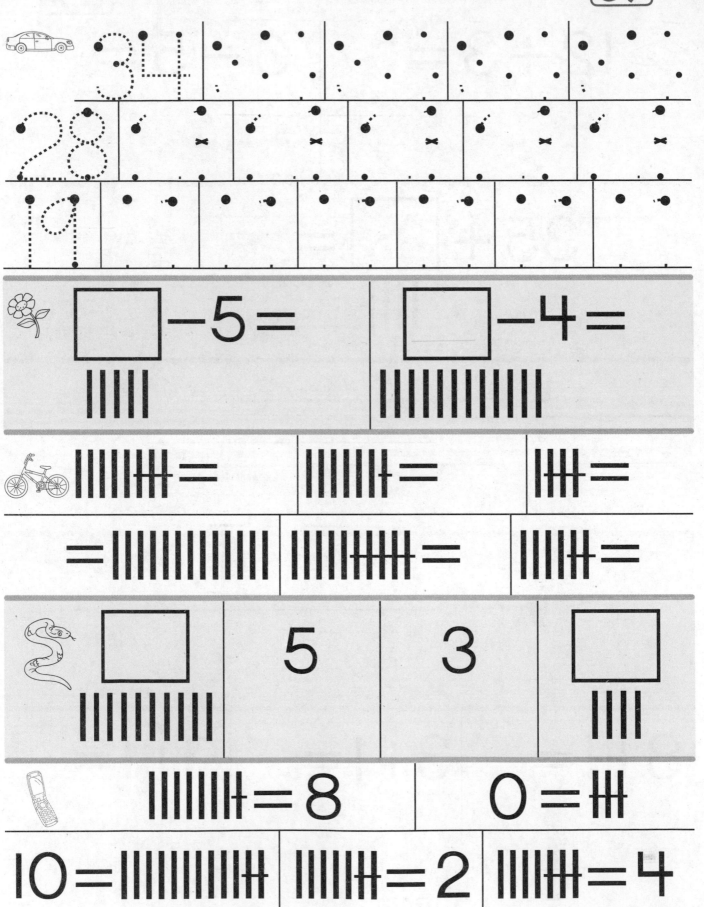

34

28

19

⬜ − 5 = |||||

⬜ − 4 = ||||||||||

|||||||| =

||||||| =

|||| =

= |||||||||||

||||||||| =

||||||| =

⬜ ||||||||| 5   3   ⬜ ||||

|||||||| = 8   0 = |||||

10 = |||||||||||| |||||| = 2 |||||| = 4

Connecting Math Concepts

$7 - 2 =$ | $8 + 2 =$

$13 +$ ▢ $=$
|||||

♡ _____

 _____

$7 + 1 =$ | $4 + 1 =$ | $6 + 1 =$

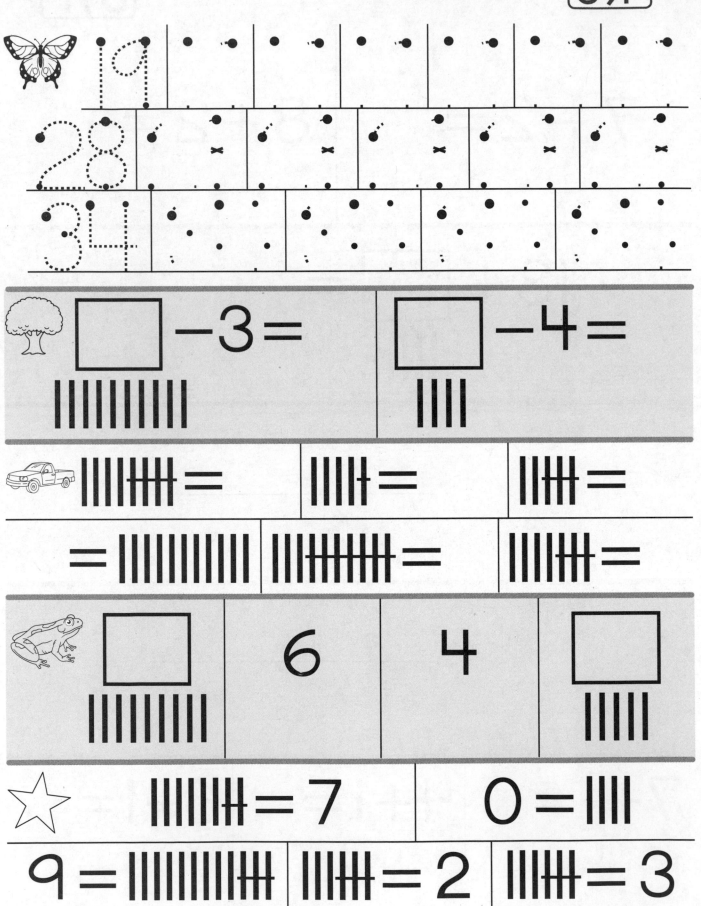

$\boxed{} - 3 =$

$\boxed{} - 4 =$

$\boxed{}$ 6 4 $\boxed{}$

|||||||| = 7   0 = ||||

9 = |||||||||||| |||||| = 2 |||||| = 3

Connecting Math Concepts

$5 + 3 =$

$7 - 4 =$ $8 - 3 =$

$14 + \boxed{} =$ $8 + \boxed{} =$
 ||||| |||||

 _____

_____

$16 + 1 =$ | $4 + 1 =$ | $7 + 1 =$

$12 + 1 =$ | $3 + 1 =$ | $9 + 1 =$

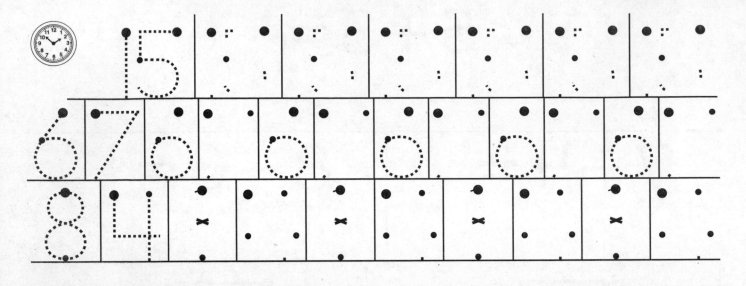

$$\boxed{} - 3 =$$

||||||||||||

$$\boxed{} - 2 =$$

||

||||||| =

|||||||||| =

= ||||||

|||||| =

||||| =

$$7 + 6 = \boxed{}$$

$$12 + 5 = \boxed{}$$

Level A Correlation to Grade K
Common Core State Standards for Mathematics

Counting and Cardinality (K.CC)

Know number names and the count sequence.

1. Count to 100 by ones and by tens.

Lessons	66–75, 80, 81

Counting and Cardinality (K.CC)

Know number names and the count sequence.

3. Write numbers from 0 to 20. Represent a number of objects with a written numeral 0–20 (with 0 representing a count of no objects).

Lessons	1–100

Counting and Cardinality (K.CC)

Count to tell the number of objects.

4. Understand the relationship between numbers and quantities; connect counting to cardinality.
   a. When counting objects, say the number names in the standard order, pairing each object with one and only one number name and each number name with one and only one object.
   b. Understand that the last number name said tells the number of objects counted. The number of objects is the same regardless of their arrangement or the order in which they were counted.
   c. Understand that each successive number name refers to a quantity that is one larger.

Lessons	15–120

Counting and Cardinality (K.CC)

Count to tell the number of objects.

5. Count to answer "how many?" questions about as many as 20 things arranged in a line, a rectangular array, or a circle, or as many as 10 things in a scattered configuration; given a number from 1–20, count out that many objects.

Lessons	15–120

Counting and Cardinality (K.CC)

Compare numbers.

6. Identify whether the number of objects in one group is greater than, less than, or equal to the number of objects in another group, e.g., by using matching and counting strategies.

Lessons	24–33, 36, 44–55, 59, 60

Operations and Algebraic Thinking (K.OA)

Understand addition as putting together and adding to, and understand subtraction as taking apart and taking from.

1. Represent addition and subtraction with objects, fingers, mental images, drawings, sounds (e.g., claps), acting out situations, verbal explanations, expressions, or equations.

Lessons	26–120

Operations and Algebraic Thinking (K.OA)

Understand addition as putting together and adding to, and understand subtraction as taking apart and taking from.

2. Solve addition and subtraction word problems, and add and subtract within 10, e.g., by using objects or drawings to represent the problem.

Lessons	36–120

Operations and Algebraic Thinking (K.OA)

Understand addition as putting together and adding to, and understand subtraction as taking apart and taking from.

3. Decompose numbers less than or equal to 10 into pairs in more than one way, e.g., by using objects or drawings, and record each decomposition by a drawing or equation (e.g., $5 = 2 + 3$ and $5 = 4 + 1$).

Lessons	76–89, 111–120

Operations and Algebraic Thinking (K.OA)

Understand addition as putting together and adding to, and understand subtraction as taking apart and taking from.

4. For any number from 1 to 9, find the number that makes 10 when added to the given number, e.g., by using objects or drawings, and record the answer with a drawing or equation.

Lessons	116–120

Operations and Algebraic Thinking (K.OA)

Understand addition as putting together and adding to, and understand subtraction as taking apart and taking from.

5. Fluently add and subtract within 5.

Lessons	56–120

Number and Operations in Base Ten (K.NBT)

Work with numbers 11–19 to gain foundations for place value.

1. Compose and decompose numbers from 11 to 19 into ten ones and some further ones, e.g., by using objects or drawings, and record each composition or decomposition by a drawing or equation (e.g., 18 = 10 + 8); understand that these numbers are composed of ten ones and one, two, three, four, five, six, seven, eight, or nine ones.

Lessons	66–93, 102–117

Measurement and Data (K.MD)

Describe and compare measurable attributes.

2. Directly compare two objects with a measurable attribute in common, to see which object has "more of"/"less of" the attribute, and describe the difference. *For example, directly compare the heights of two children and describe one child as taller/shorter.*

Measurement and Data (K.MD)

Classify objects and count the number of objects in each category.

3. Classify objects into given categories; count the number of objects in each category and sort the categories by count.

Lessons	118–120

Geometry (K.G)

Identify and describe shapes (squares, circles, triangles, rectangles, hexagons, cubes, cones, cylinders, and spheres).

1. Describe objects in the environment using names of shapes, and describe the relative positions of these objects using terms such as *above, below, beside, in front of, behind,* and *next to.*

Geometry (K.G)

Identify and describe shapes (squares, circles, triangles, rectangles, hexagons, cubes, cones, cylinders, and spheres).

2. Correctly name shapes regardless of their orientations or overall size.

Geometry (K.G)

Analyze, compare, create, and compose shapes.

5. Model shapes in the world by building shapes from components (e.g., sticks and clay balls) and drawing shapes.

Lessons	eLessons 100–101

Geometry (K.G)

Analyze, compare, create, and compose shapes.

6. Compose simple shapes to form larger shapes. *For example,"Can you join these two triangles with full sides touching to make a rectangle?"*

Lessons	eLessons 102–104

Standards for Mathematical Practice

Connecting Math Concepts addresses all of the Standards for Mathematical Practice throughout the program. What follows are examples of how individual standards are addressed in this level.

1. Make sense of problems and persevere in solving them.

Symbol Identification and Symbol Writing (Lessons 1–120): Students learn to identify and write numbers and symbols, such as plus, minus, and equals, extending that information to first translating and then solving word problems.

2. Reason abstractly and quantitatively.

Numbers and Counters (Lessons 1–120): Students learn to represent numbers and quantities with counters and apply that knowledge to adding, subtracting, and solving word problems. They learn to add and subtract using Ts and lines, where a T represents 10 and a line represents 1.

3. Construct viable arguments and critique the reasoning of others.

Shapes and 3-D Objects (Lessons 85–120): Starting in Lesson 116, students learn the difference between 2-dimensional and 3-dimensional objects, acquiring the knowledge to construct an argument for whether an object is 2-dimensional or 3-dimensional.

4. Model with mathematics.

Word Problems (Lessons 25–120): Students write and solve equations that describe situations in word problems.

5. Use appropriate tools strategically.

Throughout the program (Lessons 1–120) students use pencils and their workbooks to complete only the work that the teacher instructs them to complete.

6. Attend to precision.

Counting (Lessons 1–120): Students count objects (e.g., Ts, lines, coins) to an exact number without stopping short or going beyond. This practice is the foundation for the addition and subtraction strategies that students later learn.

7. Look for and make use of structure.

Commutative Property (Lessons 77–120): Students learn that they can "turn around" addition problems so that when they know $3 + 2 = 5$, they also know that $2 + 3 = 5$.

8. Look for and express regularity in repeated reasoning.

Column Problems (Lessons 85–120): Students add and subtract single-digit numbers by counting and representing numbers with counters (e.g., Ts and lines). They apply a parallel strategy to add and subtract two 2-digit numbers. What works for adding and subtracting basic facts also works for bigger numbers.